International Technology Roadmap for Semiconductors Lithography Nodes

Contents

Chapter 1

1 μm process

The **1 μm process** refers to the level of semiconductor process technology that was reached around the 1985-1986 timeframe[1][2] by the leading semiconductor companies, like Intel and IBM.

1.1 Products featuring 1.0 μm manufacturing process

- Intel 80386 CPU launched in 1985 was manufactured using this process.[3]

1.2 References

[1] Mueller, S (2006-07-21). "Microprocessors from 1971 to the Present". informIT. Archived from the original on 2015-04-27. Retrieved 2012-05-11.

[2] Myslewski, R (2011-11-15). "Happy 40th birthday, Intel 4004!". TheRegister. Archived from the original on 2015-04-27. Retrieved 2015-04-19.

[3] Mueller, S (2006-07-21). "Microprocessors from 1971 to the Present". informIT. Archived from the original on 2015-04-27. Retrieved 2012-05-11.

1.3 External links

- Brief timeline of microprocessor development

Chapter 2

1.5 μm process

The **1.5 μm process** is the level of semiconductor process technology that was reached around 1982 by leading semiconductor companies such as Intel and IBM.

2.1 Products featuring 1.5 μm manufacturing process

- Intel 80286 CPU launched in 1982 was manufactured using this process. [1]

2.2 References

[1] "History of the Intel Microprocessor - Listoid". Archived from the original on 2015-04-27. Retrieved 2015-04-19.

2.3 External links

- Brief timeline of microprocessor development

Chapter 3

10 μm process

The **10 μm process** is the level of semiconductor process technology that was reached around 1971[1][2] by the leading semiconductor companies such as Intel.

3.1   Products featuring 10 μm manufacturing process

- Intel 4004 CPU launched in 1971 was manufactured using this process.[3]

- Intel 8008 CPU launched in 1972 was manufactured using this process.[4]

3.2   References

[1] Mueller, S (2006-07-21). "Microprocessors from 1971 to the Present". informIT. Retrieved 2012-05-11.

[2] Myslewski, R (2011-11-15). "Happy 40th birthday, Intel 4004!". TheRegister. Archived from the original on 2015-04-27. Retrieved 2015-04-19.

[3] "History of the Intel Microprocessor - Listoid". Archived from the original on 2015-04-27. Retrieved 2015-04-19.

[4] "History of the Intel Microprocessor - Listoid". Archived from the original on 2015-04-27. Retrieved 2015-04-19.

3.3   External links

- Brief timeline of microprocessor development

Chapter 4

6 μm process

The **6 μm process** is the level of semiconductor process technology that was reached around 1974[1][2] by the leading semiconductor companies such as Intel.

4.1 Products featuring 6 μm manufacturing process

- Intel 8080 CPU launched in 1974 was manufactured using this process.[3]

4.2 References

[1] Mueller, S (2006-07-21). "Microprocessors from 1971 to the Present". informIT. Retrieved 2012-05-11.

[2] Myslewski, R (2011-11-15). "Happy 40th birthday, Intel 4004!". TheRegister.

[3] http://www.listoid.com/list/142

4.3 External links

- Brief timeline of microprocessor development

Chapter 5

10 nanometer

For the length in general and comparison, see 10 nanometres.

In semiconductor fabrication, the International Technology Roadmap for Semiconductors (ITRS) defines the **10 nanometer** (**10 nm**) node as the technology node following the 14 nm node. "**10 nm class**" denotes chips made using process technologies between 10 and 20 nanometers.

As of 2016, 10 nm devices are still under commercial development.

5.1 History

5.1.1 Background

The ITRS' original naming of this technology node was "11 nm". According to the 2007 edition of the roadmap, by the year 2022, the half-pitch (i.e., half the distance between identical features in an array) for a DRAM should be 11 nm. Pat Gelsinger, at the time serving as Intel's Chief Technology Officer, claimed in 2008 that Intel saw a 'clear way' towards the 10 nm node.[1][2] At the 11 nm node, Intel expected (in 2006) to be using a half-pitch of around 21 nm, in 2015,[3] Nvidia's chief scientist, William Dally, claimed (in 2009) that they would also reach 11 nm semiconductors in 2015, a transition he claimed would be facilitated principally through new electronic design automation tools.[4]

This 10 nm design rule is considered likely to be realized by multiple patterning,[5][6][7] given the difficulty of implementing EUV lithography.

5.1.2 Potential technologies

While the roadmap has been based on the continuing extension of CMOS technology, even this roadmap does not guarantee that silicon-based CMOS will extend that far. This is to be expected, since the gate length for this node may be smaller than 6 nm, and the corresponding gate dielectric thickness would scale down to a monolayer or even less. Scientists have estimated that transistors at these dimensions are significantly affected by quantum tunnelling.[8] As a result, non-silicon extensions of CMOS, using III-V materials or Carbon nanotube/nanowires, as well as non-CMOS platforms, including molecular electronics, spin-based computing, and single-electron devices, have been proposed. Hence, this node marks the practical beginning of nanoelectronics.

The extensive use of ultra-low-k dielectrics (such as spin-on polymers or other porous materials) means that conventional photolithography, etch, or even chemical-mechanical polishing processes are unlikely to be used, because these materials contain a high density of voids and gaps. At the ~10 nm scale, quantum tunneling (especially through gaps) becomes a significant phenomenon.[9] Controlling gaps on these scales by means of electromigration can produce interesting electrical properties.[10]

Quantum tunneling may not be a disadvantage if its effect on device behavior can be understood, and exploited, in the design. Future transistors may have insulating channels. An electron wave function decays exponentially in a "classically forbidden" region at a rate that can be controlled by the gate voltage. Interference effects are also possible;[11] Alternate option is in heavier mass semiconducting channels.[12] Photoemission electron microscopy (PEEM) data has been used to show that low energy electrons ~1.35 eV could travel as far as ~15 nm in SiO_2, despite an average measured attenuation length of 1.18 nm.[13]

5.1.3   Technology demos

In April 2015, TSMC announced that 10 nm production would begin at the end of 2016.[14]

On 23 May 2015, Samsung Electronics showed off a 300 mm wafer of 10 nm FinFET chips.[15]

5.1.4   Mass production

As of January 2016, none.

5.2   References

[1] Damon Poeter. "Intel's Gelsinger Sees Clear Path To 10nm Chips". Archived from the original on 2009-06-22. Retrieved 2009-06-20.

[2] "MIT: Optical lithography good to 12 nanometers". Archived from the original on 2009-06-22. Retrieved 2009-06-20.

[3] Borodovsky, Y. (2006). "Marching to the beat of Moore's Law". *Proc. SPIE* **6153**. doi:10.1117/12.655176.

[4] "Nvidia Chief Scientist: 11nm Graphics Chips with 5000 Stream Processors Due in 2015". XBit Labs. July 30, 2009. Archived from the original on 2009-09-03. Retrieved 2009-08-27.

[5] SEMICON West - Lithography Challenges and Solutions

[6] J. Word *et al.*, Proc. SPIE 6925 (2008).

[7] Intel extending ArF lithography

[8] "Intel scientists find wall for Moore's Law". ZDNet. December 1, 2003.

[9] Naitoh, Y.; et al. (2007). "New Nonvolatile Memory Effect Showing Reproducible Large Resistance Ratio Employing Nanogap Gold Junction". *MRS Symposium Proceedings* **997**: 0997–I04–08. doi:10.1557/PROC-0997-I04-08.

[10] Kayashima, S.; et al. (2007). "Control of Tunnel Resistance of Nanogaps by Field-Emission-Induced Electromigration". *Jap. J. Appl. Phys.* **46** (36–40): L907–909. doi:10.1143/JJAP.46.L907.

[11] Ahmed, Khaled; Schuegraf, Klaus (November 2011). "Transistor Wars: Rival architectures face off in a bid to keep Moore's Law alive". *IEEE Spectrum*: 50.

[12] Mehrotra, S.; et al. (2013). "Engineering Nanowire n-MOSFETs at Lg < 8 nm". *Preprint*. arXiv:1303.5458.

[13] Ballarotto, V. W.; et al. (2002). "Photoelectron emission microscopy of ultrathin oxide covered devices". *JVST B* **20** (6): 2514–2518. doi:10.1116/1.1525007.

[14] "TSMC Launching 10 nm FinFET Process In 2016, 7nm In 2017". 19 April 2015. Retrieved 25 May 2015.

[15] "Samsung vows to start 10nm chip production in 2016". 23 May 2015. Retrieved 16 July 2015.

Chapter 6

130 nanometer

The **130 nanometer** (**130 nm**) process refers to the level of semiconductor process technology that was reached in the 2000–2001 timeframe, by most leading semiconductor companies, like Intel, Texas Instruments, IBM, and TSMC.

The origin of the 130 nm value is historical, as it reflects a trend of 70% scaling every 2–3 years. The naming is formally determined by the International Technology Roadmap for Semiconductors (ITRS).

Some of the first CPUs manufactured with this process include Intel Tualatin family of Pentium III processors.

6.1 Processors using 130 nm manufacturing technology

- Motorola PowerPC 7447 and 7457 2002

- IBM Gekko (Nintendo GameCube)

- IBM PowerPC G5 970 - October 2002 - June 2003

- Intel Pentium III Tualatin and Coppermine - 2001-04

- Intel Celeron Tualatin−256 - 2001-10-02

- Intel Pentium M Banias - 2003-03-12

- Intel Pentium 4 Northwood- 2002-01-07

- Intel Celeron Northwood-128 - 2002-09-18

- Intel Xeon Prestonia and Gallatin - 2002-02-25

- VIA C3 - 2001

- AMD Athlon XP Thoroughbred, Thorton, and Barton

- AMD Athlon MP Thoroughbred - 2002-08-27

- AMD Athlon XP-M Thoroughbred, Barton, and Dublin

- AMD Duron Applebred - 2003-08-21

- AMD K7 Sempron Thoroughbred-B, Thorton, and Barton - 2004-07-28

- AMD K8 Sempron Paris - 2004-07-28

- AMD Athlon 64 Clawhammer and Newcastle - 2003-09-23

- AMD Opteron Sledgehammer - 2003-06-30

- Elbrus 2000 1891ВМ4Я (1891VM4YA) - 2008-04-27 [1]

- MCST-R500S 1891BM3 - 2008-07-27 [2]

- Vortex 86SX - [3]

6.2   References

[1] (Russian) "Микропроцессор Эльбрус/МЦСТ". *Mcst.ru*. Retrieved 2015-09-10.

[2] (Russian) "Микропроцессор МЦСТ R500S/МЦСТ". *Mcst.ru*. Retrieved 2015-09-10.

[3] "CPU from DM&P". *Dmp.com.tw*. Retrieved 2015-09-10.

Chapter 7

14 nanometer

Both "14nm" and "16nm" fabrication nodes are discussed here

The **14 nanometer** (**14 nm**) semiconductor device fabrication node is the technology node following the 22 nm/(20 nm) node. The naming of this technology node as "14 nm" came from the International Technology Roadmap for Semiconductors (ITRS).

The first 14 nm scale devices were shipped to consumers by Intel in 2014.

7.1 History

7.1.1 Background

14 nm resolution is difficult to achieve in a polymeric resist, even with electron beam lithography. In addition, the chemical effects of ionizing radiation also limit reliable resolution to about 30 nm, which is also achievable using current state-of-the-art immersion lithography. Hardmask materials and multiple patterning are required.

A more significant limitation comes from plasma damage to low-k materials. The extent of damage is typically 20 nm thick,[1] but can also go up to about 100 nm.[2] The damage sensitivity is expected to get worse as the low-k materials become more porous.

For comparison, the atomic radius of unstrained silicon is 111 pm (0.111 nm). Thus about 90 Si atoms would span the channel length, leading to substantial leakage.

Tela Innovations and Sequoia Design Systems developed a methodology allowing double exposure for the 14 nm node. c.2010.[3]

Samsung and Synopsys have also begun implementing double patterning in 22 nm and 16 nm design flows.[4]

Mentor Graphics reported taping out 16 nm test chips in 2010.[5]

On January 17, 2011, IBM announced that they were teaming up with ARM to develop 14 nm chip processing technology.[6]

On February 18, 2011, Intel announced that it would construct a new $5 billion semiconductor fabrication plant in Arizona, designed to manufacture chips using the 14 nm manufacturing processes and leading-edge 300 mm wafers.[7] The new fabrication plant was to be named Fab 42, and construction was meant to start in the middle of 2011. Intel billed the new facility as "the most advanced, high-volume manufacturing facility in the world," and said it would come on line in 2013. Intel has since decided to postpone opening this facility and instead upgrade its existing facilities to support 14-nm chips.[8] On May 17, 2011, Intel announced a roadmap for 2014 that included 14 nm transistors for their Xeon, Core, and Atom product lines.[9]

7.1.2   Technology demos

In 2005, Toshiba demonstrated 15 nm gate length and 10 nm fin width using a sidewall spacer process.[10] It has been suggested that for the 16 nm node, a logic transistor would have a gate length of about 5 nm.[11] In December 2007, Toshiba demonstrated a prototype memory unit that used 15 nanometer thin lines.[12]

In December 2009, National Nano Device Laboratories, owned by the Taiwanese government, produced a 16 nm SRAM chip.[13]

In September 2011, Hynix announced the development of 15 nm NAND cells.[14]

In December 2012, Samsung Electronics taped out a 14 nm chip.[15]

In September 2013, Intel demonstrated an Ultrabook laptop that used a 14 nm Broadwell CPU and Intel CEO Brian Krzanich said "[CPU] will be shipping by the end of this year."[16] However, shipment had been delayed further until Q4 2014.[17]

In August 2014, Intel announced details of the 14 nm microarchitecture for its upcoming Core M processors, the first product to be manufactured on Intel's 14 nm manufacturing process. The first systems based on the Core M processor were to become available in Q4 2014 - according to the press release. "Intel's 14 nanometer technology uses second-generation Tri-gate transistors to deliver industry-leading performance, power, density and cost per transistor," said Mark Bohr, Intel senior fellow, Technology and Manufacturing Group, and director, Process Architecture and Integration.[18]

7.1.3   Shipping devices

On 5 September 2014, Intel launched the first three Broadwell-based processors that belonged to the low-TDP Core M family, Core M 5Y10, Core M 5Y10a and Core M 5Y70.[19]

In February 2015, Samsung announced its flagship smartphones Galaxy S6 and Galaxy S6 Edge would feature 14 nm Exynos systems-on-a-chip.[20]

On March 9, 2015, Apple Inc. released the "Early 2015" MacBook and MacBook Pro, which utilized 14 nm Intel processors. Of note is the i7-5557U, which has Intel Iris 6100 graphics and two cores running at 3.1Ghz, using only 28 watts.[21][22]

On September 25, 2015, Apple Inc. released iPhone 6s and iPhone 6s Plus, which equipped with "desktop-class" A9 chips[23] that are fabricated in both 14 nm by Samsung and 16 nm by TSMC.

7.2   References

[1] Richard, O.; et al. (2007). "Sidewall damage in silica-based low-*k* material induced by different patterning plasma processes studied by energy filtered and analytical scanning TEM". *Microelectronic Engineering* **84** (3): 517–523. doi:10.1016/j.mee.2006.10.058.

[2] Gross, T.; et al. (2008). "Detection of nanoscale etch and ash damage to nanoporous methyl silsesquioxane using electrostatic force microscopy". *Microelectronic Engineering* **85** (2): 401–407. doi:10.1016/j.mee.2007.07.014.

[3] Axelrad, V.; et al. (2010). "16nm with 193nm immersion lithography and double exposure". *Proc. SPIE* **7641**: 764109. doi:10.1117/12.846677.

[4] Noh, M-S.; et al. (2010). "Implementing and validating double patterning in 22-nm to 16-nm product design and patterning flows". *Proc. SPIE* **7640**: 76400S. doi:10.1117/12.848194.

[5] "Mentor moves tools toward 16-nanometer". EETimes. August 23, 2010.

[6] "IBM and ARM to Collaborate on Advanced Semiconductor Technology for Mobile Electronics". *IBM Press release*. January 17, 2011.

[7] "Intel to build fab for 14-nm chips". EE Times.

[8] "Intel shelves cutting-edge Arizona chip factory". *Reuters*. January 14, 2014.

[9] "Implementing and validating double patterning in 22-nm to 16-nm product design and patterning flows". *AnandTech*. May 17, 2011.

[10] Kaneko, A; Yagashita, A; Yahashi, K; Kubota, T; et al. (2005). "Sidewall transfer process and selective gate sidewall spacer formation technology for sub-15nm FinFET with elevated source/drain extension". *IEEE International Electron Devices Meeting (IEDM 2005)*. pp. 844–847. doi:10.1109/IEDM.2005.1609488.

[11] "Intel scientists find wall for Moore's Law". ZDNet. December 1, 2003.

[12] "15 Nanometre Memory Tested". *The Inquirer*.

[13] "16nm SRAM produced – Taiwan Today". taiwantoday.tw.

[14] Hübler, Arved; et al. (2011). "Printed Paper Photovoltaic Cells". *Advanced Energy Materials* **1** (6): 1018–1022. doi:10.1002/aenm.201100394.

[15] "Samsung reveals its first 14nm FinFET test chip". Engadget. December 21, 2012.

[16] "Intel reveals 14nm PC, declares Moore's Law 'alive and well'". The Register. September 10, 2013.

[17] "Intel postpones Broadwell availability to 4Q14". Digitimes.com. Retrieved 2014-02-13.

[18] "Intel Discloses Newest Microarchitecture and 14 Nanometer Manufacturing Process Technical Details". Intel. August 11, 2014.

[19] Shvets, Anthony (7 September 2014). "Intel launches first Broadwell processors". *CPU World*. Retrieved 18 March 2015.

[20] Samsung Announces Mass Production of Industry's First 14nm FinFET Mobile Application Processor

[21] "Apple MacBook Pro "Core i7" 3.1 13" Early 2015 Specs". *EveryMac.com*. 2015. Retrieved 18 March 2015.

[22] "Intel Core i7-5557U specifications". *CPU World*. 2015. Retrieved 18 March 2015.

[23] "Apple's new A9 and A9X processors promise 'desktop-class performance'". *The Verge*. 2015. Retrieved 11 October 2015.

Chapter 8

180 nanometer

The **180 nanometer** (**180 nm**) process refers to the level of semiconductor process technology that was reached in the 1999-2000 timeframe by most leading semiconductor companies, like Intel, Texas Instruments, IBM, and TSMC.

The origin of the 180 nm value is historical, as it reflects a trend of 70% scaling every 2–3 years. The naming is formally determined by the International Technology Roadmap for Semiconductors (ITRS).

Some of the first CPUs manufactured with this process include Intel Coppermine family of Pentium III processors. This was the first technology using a gate length shorter than that of light used for lithography (which has a minimum of 193 nm).

Some more recent microprocessors and microcontrollers (e.g. PIC) are using this technology because it is typically low cost and does not require upgrading of existing equipment.

8.1   Processors using 180 nm manufacturing technology

- Intel Coppermine E- October, 1999
- AMD Athlon Thunderbird - June 2000
- Intel Celeron (Willamette) - May, 2002
- Motorola PowerPC 7445 and 7455 (Apollo 6) - January, 2002

Chapter 9

22 nanometer

The **22 nanometer** (**22 nm**) node is the process step following the 32 nm in CMOS semiconductor device fabrication. The typical half-pitch (i.e., half the distance between identical features in an array) for a memory cell using the process is around 22 nm. It was first introduced by semiconductor companies in 2008 for use in memory products, while first consumer-level CPU deliveries started in April 2012.

The ITRS 2006 Front End Process Update indicates that equivalent physical oxide thickness will not scale below 0.5 nm (about twice the diameter of a silicon atom), which is the expected value at the 22 nm node. This is an indication that CMOS scaling in this area has reached a wall at this point, possibly disturbing Moore's law.

The 22 nm process was superseded by commercial 14 nm technology in 2014.

9.1 Technology demos

On August 18, 2008, AMD, Freescale, IBM, STMicroelectronics, Toshiba, and the College of Nanoscale Science and Engineering (CNSE) announced that they jointly developed and manufactured a 22 nm SRAM cell, built on a traditional six-transistor design on a 300 mm wafer, which had a memory cell size of just 0.1 μm^2.[1] The cell was printed using immersion lithography.[2]

The 22 nm node may be the first time where the gate length is not necessarily smaller than the technology node designation. For example, a 25 nm gate length would be typical for the 22 nm node.

On September 22, 2009, during the Intel Developer Forum Fall 2009, Intel showed a 22 nm wafer and announced that chips with 22 nm technology would be available in the second half of 2011.[3] SRAM cell size is said to be 0.092 μm^2, smallest reported to date.

On January 3, 2010, Intel and Micron Technology announced the first in a family of 25 nm NAND devices.

On May 2, 2011, Intel announced its first 22 nm microprocessor, codenamed Ivy Bridge, using a technology called 3-D Tri-Gate.[4]

POWER8 processors is produced in a 22 nm SOI process.[5]

9.2 Shipping devices

On August 31, 2010, Toshiba announced that it was shipping 24 nm flash memory NAND devices.[6]

In 2010, Hynix Semiconductor announced that it has used a 26 nm manufacturing process to produce a flash device with 64 Gbit capacity; Intel Corp. and Micron Technology had by then already developed the technology themselves.[7]

On April 23, 2012, Intel Core i7 and Intel Core i5 processors based on Intel's Ivy Bridge 22 nm technology for series

7 chipsets went on sale worldwide.[8] Volume production of 22 nm processors began more than six months earlier, as confirmed by former Intel CEO Paul Otellini on October 19, 2011.[9]

On June 3, 2013, Intel started shipping Intel Core i7 and Intel Core i5 processors based on Intel's Haswell microarchitecture in 22 nm technology for series 8 chipsets.[10]

9.3   References

[1] TG Daily news report

[2] EETimes news report

[3] Intel announces 22nm chips for 2011

[4] Intel 22nm 3-D Tri-Gate Transistor Technology

[5] IBM opens Power8 kimono (a little bit more)

[6] Toshiba launches 24nm process NAND flash memory

[7] Article reporting Hynix 26 nm technology announcement

[8] Intel launches Ivy Bridge...

[9] Tom's Hardware: Intel to Sell Ivy Bridge Late in Q4 2011

[10] 4th Generation Intel® Core™ Processors Coming Soon

Chapter 10

250 nanometer

The **250 nanometer** (**250 nm**) process refers to a level of semiconductor process technology that was reached by most manufacturers in the 1997-1998 timeframe.

10.1   Products featuring 250 nm manufacturing process

- The DEC Alpha 21264A, which was made commercially available in 1999.

- The AMD K6-2 *Chomper* and *Chomper Extended*. Chomper was released on May 28, 1998.

- The AMD K6-III "Sharptooth" used 250 nm.

- The mobile Pentium MMX *Tillamook*, released in August 1997.

- The Pentium II *Deschutes*.

- The Pentium III *Katmai*.

- The Dreamcast CPU and GPU.

- The initial PlayStation 2's Emotion Engine CPU.

Chapter 11

3 μm process

The **3 μm process** is the level of semiconductor process technology that was reached in 1977[1][2] by the leading semiconductor companies such as Intel.

11.1 Products featuring 3 μm manufacturing process

- Intel 8085 CPU launched in 1977 was manufactured using 3.2 μm process.[3]

- Intel 8086 CPU launched in 1978 was manufactured using 3.2 μm process.[4]

- Intel 8088 CPU launched in 1979 was manufactured using 3.2 μm process.[5]

11.2 References

[1] Mueller, S (2006-07-21). "Microprocessors from 1971 to the Present". informIT. Retrieved 2012-05-11.

[2] Myslewski, R (2011-11-15). "Happy 40th birthday, Intel 4004!". TheRegister.

[3] Mueller, S (2006-07-21). "Microprocessors from 1971 to the Present". informIT. Retrieved 2012-05-11.

[4] http://www.listoid.com/list/142

[5] http://www.listoid.com/list/142

Chapter 12

32 nanometer

The **32 nanometer** (**32 nm**) node is the step following the 45 nanometer process in CMOS semiconductor device fabrication. "32 nanometer" refers to the average half-pitch (i.e., half the distance between identical features) of a memory cell at this technology level. Intel and AMD both produced commercial microchips using the 32 nanometer process in the early 2010s. IBM and the Common Platform also developed a 32 nm high-k metal gate process.[1] Intel began selling its first 32 nm processors using the Westmere architecture on 7 January 2010. The 32 nm process was superseded by commercial 22 nm technology in 2012.[2][3]

12.1 Technology demos

Prototypes using 32 nm technology first emerged in the mid-2000s. In 2004, IBM demonstrated a 0.143 μm^2 SRAM cell with a poly gate pitch of 135 nm, produced using electron-beam lithography and photolithography on the same layer. It was observed that the cell's sensitivity to input voltage fluctuations degraded significantly at such a small scale.[4] In October 2006, the Interuniversity Microelectronics Centre (IMEC) demonstrated a 32 nm flash patterning capability based on double patterning and immersion lithography.[5] The necessity of introducing double patterning and hyper-NA tools to reduce memory cell area offset some of the cost advantages of moving to this node from the 45 nm node.[6] TSMC similarly used double patterning combined with immersion lithography to produce a 32 nm node 0.183 μm^2 six-transistor SRAM cell in 2005.[7]

Intel Corporation revealed its first 32 nm test chips to the public on 18 September 2007 at the Intel Developer Forum. The test chips had a cell size of 0.182 μm^2, used a second-generation high-k gate dielectric and metal gate, and contained almost two billion transistors. 193 nm immersion lithography was used for the critical layers, while 193 nm or 248 nm dry lithography was used on less critical layers. The critical pitch was 112.5 nm.[8]

In January 2011, Samsung completed development of what it claimed was the industry's first DDR4 DRAM module using a process technology with a size between 30 nm and 39 nm. The module could reportedly achieve data transfer rates of 2.133 Gbit/s at 1.2V, compared to 1.35V and 1.5V DDR3 DRAM at an equivalent 30 nm-class process technology with speeds of up to 1.6 Gbit/s. The module used pseudo open drain (POD) technology, specially adapted to allow DDR4 DRAM to consume just half the current of DDR3 when reading and writing data.[9]

12.2 Processors using 32 nm technology

Intel's Core i3 and i5 processors, released in January 2010, were among the first mass-produced processors to use 32 nm technology.[10] Intel's second-generation Core processors, codenamed Sandy Bridge, also used the 32 nm manufacturing process. Intel's 6-core processor, codenamed Gulftown and built on the Westmere architecture, was released on 16 March 2010 as the Core i7 980x Extreme Edition, retailing for approximately US$1,000.[11] Intel's lower-end 6-core, the i7-970, was released in late July 2010, priced at approximately US$900.

17

AMD also released 32 nm SOI processors in the early 2010s. AMD's FX Series processors, codenamed Zambezi and based on AMD's Bulldozer architecture, were released in October 2011. The technology utilised a 32 nm SOI process, two CPU cores per module, and up to four modules, ranging from a quad-core design costing approximately US$130 to a $280 eight-core design.

In September 2011, Ambarella Inc. announced the availability of the 32 nm-based A7L system-on-a-chip circuit for digital still cameras, providing 1080p60 high-definition video capabilities.[12]

12.3   Successor node

The successor to 32 nm technology was the 22 nm node, per the International Technology Roadmap for Semiconductors. Intel began mass production of 22 nm semiconductors in late 2011,[13] and announced the release of its first commercial 22 nm devices in April 2012.[2][14]

12.4   References

[1] Intel (Architecture & Silicon). Gate Dielectric Scaling for CMOS: from SiO_2/PolySi to High-K/Metal-Gate. White Paper. Intel.com. Retrieved 18 June 2013.

[2] "Report: Intel Scheduling 22 nm Ivy Bridge for April 2012". Tom'sHardware.com. 26 November 2011. Retrieved 5 December 2011.

[3] "Intel's Ivy Bridge chips launch using '3D transistors'". BBC. 23 April 2012. Retrieved 18 June 2013.

[4] D. M. Fried et al., IEDM 2004.

[5] "IMEC demonstrates feasibility of double patterning immersion litho for 32nm node". PhysOrg.com. 18 October 2006. Retrieved 17 December 2011.

[6] Mark LaPedus (23 February 2007). "IBM sees immersion at 22nm, pushes out EUV". *EE Times*. Retrieved 11 November 2011.

[7] H-Y. Chen et al., Symp. on VLSI Tech. 2005.

[8] F. T. Chen (2002). *Proc. SPIE*. Vol. 4889, no. 1313.

[9] Peter Clarke (4 January 2011). "Samsung trials DDR4 DRAM module". *EE Times*. Retrieved 11 November 2011.

[10] "Intel Debuts 32-NM Westmere Desktop Processors". *Information Week*. 7 January 2010. Retrieved 17 December 2011.

[11] Sal Cangeloso (4 February 2010). "Intel's 6-core 32nm processors arriving soon". Geek.com. Retrieved 11 November 2011.

[12] "Ambarella A7L Enables the Next Generation of Digital Still Cameras with 1080p60 Fluid Motion Video". Ambarella.com. 26 September 2011. Retrieved 11 November 2011.

[13] "Intel's CEO Discusses Q3 2011 Results - Earnings Call Transcript". Seeking Alpha. 18 October 2011. Retrieved 14 February 2013.

[14] "Intel beats analysts' first quarter forecasts". BBC. 17 April 2012. Retrieved 18 June 2013.

12.5   Further reading

- Steen, S.; et al. (2006). "Hybrid lithography: The marriage between optical and e-beam lithography. A method to study process integration and device performance for advanced device nodes". *Microelec. Eng.* **83** (4–9): 754–761. doi:10.1016/j.mee.2006.01.181.

12.6 External links

- Chipmakers gear up for manufacturing hurdles

- Sony, IBM, and Toshiba partnering on semiconductor research

- IBM and AMD partnering on semiconductor research

- Slashdot discussion

- Intel 32 nm process

- [http://sst.pennnet.com/display_article/309943/5/ARTCL/none/none/1/Samsung-touts-30 nm-NAND-flash-using-double-patterning/ Samsung self-aligned double patterning technology]

Chapter 13

350 nanometer

The **350 nanometer** (**350 nm**) process refers to the level of semiconductor process technology that was reached in the 1995–1996 timeframe, by most leading semiconductor companies, like Intel and IBM.

13.1 Products featuring 350 nm manufacturing process

- Intel Pentium Pro (1995), Pentium (P54CS, 1995), and initial Pentium II CPUs (Klamath, 1997).
- AMD K5 (1996) and original AMD K6 (Model 6, 1997) CPUs.
- NEC VR4300, used in the Nintendo 64 game console.
- Parallax Propeller, 8 core microcontroller[1]

13.2 References

[1] "Propeller I semiconductor process technology? Is it 350nm or 180nm? - Parallax Forums". Forums.parallax.com. Retrieved 2015-09-13.

Chapter 14

45 nanometer

Per the International Technology Roadmap for Semiconductors, the **45 nanometer** (**45 nm**) technology node should refer to the average half-pitch of a memory cell manufactured at around the 2007–2008 time frame.

Matsushita and Intel started mass-producing 45 nm chips in late 2007, and AMD started production of 45 nm chips in late 2008, while IBM, Infineon, Samsung, and Chartered Semiconductor have already completed a common 45 nm process platform. At the end of 2008, SMIC was the first China-based semiconductor company to move to 45 nm, having licensed the bulk 45 nm process from IBM.

Many critical feature sizes are smaller than the wavelength of light used for lithography (i.e., 193 nm and 248 nm). A variety of techniques, such as larger lenses, are used to make sub-wavelength features. Double patterning has also been introduced to assist in shrinking distances between features, especially if dry lithography is used. It is expected that more layers will be patterned with 193 nm wavelength at the 45 nm node. Moving previously loose layers (such as Metal 4 and Metal 5) from 248 nm to 193 nm wavelength is expected to continue, which will likely further drive costs upward, due to difficulties with 193 nm photoresists.

14.1   High-k dielectrics

Chipmakers have initially voiced concerns about introducing new high-k materials into the gate stack, for the purpose of reducing leakage current density. As of 2007, however, both IBM and Intel have announced that they have high-k dielectric and metal gate solutions, which Intel considers to be a fundamental change in transistor design.[1] NEC has also put high-k materials into production.

14.2   Technology demos

- In 2004, TSMC demonstrated a 0.296 square micrometer 45 nm SRAM cell. In 2008, TSMC moved on to a 40 nm process.

- In January 2006, Intel demonstrated a 0.346 square micrometers 45 nm node SRAM cell.

- In April 2006, AMD demonstrated a 0.370 square micrometer 45 nm SRAM cell.

- In June 2006, Texas Instruments debuted a 0.24 square micrometer 45 nm SRAM cell, with the help of immersion lithography.

- In November 2006, UMC announced that it had developed a 45 nm SRAM chip with a cell size of less than 0.25 square micrometer using immersion lithography and low-k dielectrics.

- In June 2007 Matsushita Electric Industrial Co. started mass production of System-on-a-chip (SoC) for use in digital consumer equipment based on the 45-nm process technology.

The successors to 45 nm technology are 32 nm, 22 nm, and then 14 nm technologies.

14.3   Commercial introduction

Matsushita Electric Industrial Co. started mass production of System-on-a-chip (SoC) for use in digital consumer equipment based on the 45-nm process technology.

Intel shipped its first 45 nanometer based processor, the Xeon 5400-series, in November 2007.

Many details about Penryn appeared at the April 2007 Intel Developer Forum. Its successor is called Nehalem. Important advances[2] include the addition of new instructions (including SSE4, also known as Penryn New Instructions) and new fabrication materials (most significantly a hafnium-based dielectric).

AMD released its Sempron II, Athlon II, Turion II and Phenom II (in generally increasing order of strength), as well as Shanghai Opteron processors using the 45-nm process technology.

The Xbox 360 S, released in 2010, has its **Xenon** processor in 45 nm process.[3]

The PlayStation 3 Slim model introduced Cell Broadband Engine in 45 nm process.[4]

14.4   Example: Intel's 45 nm process

At IEDM 2007, more technical details of Intel's 45 nm process were revealed.[5]

Since immersion lithography is not used here, the lithographic patterning is more difficult. Hence many lines have been lengthened rather than shortened. A more time-consuming double patterning method is used explicitly for this 45 nm process, resulting in potentially higher risk of product delays than before. Also, the use of high-k dielectrics is introduced for the first time, to address gate leakage issues. For the 32 nm node, immersion lithography will begin to be used by Intel.

- 160 nm gate pitch (73% of 65 nm generation)

- 200 nm isolation pitch (91% of 65 nm generation) indicating a slowing of scaling of isolation distance between transistors

- Extensive use of dummy copper metal and dummy gates[6]

- 35 nm gate length (same as 65 nm generation)

- 1 nm equivalent oxide thickness, with 0.7 nm transition layer

- Gate-last process using dummy polysilicon and damascene metal gate

- Squaring of gate ends using a second photoresist coating[7]

- 9 layers of carbon-doped oxide and Cu interconnect, the last being a thick "redistribution" layer

- Contacts shaped more like rectangles than circles for local interconnection

- Lead-free packaging

- 1.36 mA/um nFET drive current

- 1.07 mA/um pFET drive current, 51% faster than 65 nm generation, with higher hole mobility due to increase from 23% to 30% Ge in embedded SiGe stressors

In a recent Chipworks reverse-engineering,[8] it was disclosed that the trench contacts were formed as a "Metal-0" layer in tungsten serving as a local interconnect. Most trench contacts were short lines oriented parallel to the gates covering diffusion, while gate contacts where even shorter lines oriented perpendicular to the gates.

It was recently revealed[9] that both the Nehalem and Atom microprocessors used SRAM cells containing eight transistors instead of the conventional six, in order to better accommodate voltage scaling. This resulted in an area penalty of over 30%.

14.5 Processors using 45 nm technology

- Matsushita has released the 45 nm Uniphier.

- Wolfdale, Wolfdale-3M, Yorkfield, Yorkfield XE and Penryn where Intel cores sold under the Core 2 brand.

- Intel Core i7 series processors, i5 750 (Lynnfield and Clarksfield).

- Pentium Dual-Core Wolfdale-3M are current Intel mainstream dual core sold under the Pentium brand.

- Diamondville, Pineview are current Intel cores with Hyper-Threading sold under the Intel Atom brand.

- AMD Deneb (Phenom II) and Shanghai (Opteron) Quad-Core Processors, Regor (Athlon II) dual core processors, Caspian (Turion II) mobile dual core processors.

- AMD (Phenom II) "Thuban" Six-Core Processor (1055T)

- **Xenon** on Xbox 360 S model.

- Cell Broadband Engine in PlayStation 3 Slim model – September 2009.

- Samsung S5PC110, as known as *Hummingbird*.

- Texas Instruments OMAP 3 and 4 series.

- IBM POWER7 and z196

- Fujitsu SPARC64 VIIIfx series

- The Wii U "Espresso" IBM CPU.

14.6 References

[1] IEEE Spectrum: The High-k Solution

[2] "Report on Penryn Series Improvements." (PDF). Intel. October 2006.

[3] "New Xbox 360 gets official at $299, shipping today, looks angular and ominous (video hands-on!)". AOL Engadget. 14 June 2010. Archived from the original on 17 June 2010. Retrieved 11 July 2010..

[4] "Sony answears our questions about the new PlayStation 3". Ars Technica. 18 August 2009. Retrieved 19 August 2009..

[5] Mistry, K.; Allen, C. and Auth, C. and Beattie, B. and Bergstrom, D. and Bost, M. and Brazier, M. and Buehler, M. and Cappellani, A. and Chau, R. and Choi, C.-H. and Ding, G. and Fischer, K. and Ghani, T. and Grover, R. and Han, W. and Hanken, D. and Hattendorf, M. and He, J. and Hicks, J. and Huessner, R. and Ingerly, D. and Jain, P. and James, R. and Jong, L. and Joshi, S. and Kenyon, C. and Kuhn, K. and Lee, K. and Liu, H. and Maiz, J. and McIntyre, B. and Moon, P. and Neirynck, J. and Pae, S. and Parker, C. and Parsons, D. and Prasad, C. and Pipes, L. and Prince, M. and Ranade, P. and Reynolds, T. and Sandford, J. and Shifren, L. and Sebastian, J. and Seiple, J. and Simon, D. and Sivakumar, S. and Smith, P. and Thomas, C. and Troeger, T. and Vandervoorn, P. and Williams, S. and Zawadzki, K. (December 2007). "A 45nm Logic Technology with High-k+Metal Gate Transistors, Strained Silicon, 9 Cu Interconnect Layers, 193nm Dry Patterning, and 100% Pb-free Packaging". doi:10.1109/IEDM.2007.4418914.

[6] http://www.ipfrontline.com/depts/article.asp?id=19560&deptid=5

[7] Intel 45 nm process at IEDM

[8] analysis

[9] 8T SRAM used for Nehalem and Atom

14.7   External links

- Panasonic Begins Mass Production of 45-nm Generation SoC

- Intel 45 nm process is good to go

- Intel moving to 45nm sooner than expected?

- Chipmakers gear up for manufacturing hurdles

- Intel 45 nm node SRAM cell

- An AMD Update

- Slashdot discussion of n nm process naming

- 45 nm Technology from Intel

- Intel 45 nm process at IEDM

Chapter 15

5 nanometer

In semiconductor manufacturing, the International Technology Roadmap for Semiconductors defines the **5 nanometer** (**5 nm**) node as the technology node following the 7 nm node.

Transistors at the 7 nm scale were first produced by researchers in the first decade of the 21st century – the process scale may represent the end of Moore's Law scaling for electronic devices.

As of 2016, no 5 nm scale devices have been commercially produced.

15.1   History

15.1.1   Background

The 5 nm node was once assumed by some experts to be the end of Moore's law.[1] Transistors smaller than 7 nm will experience quantum tunnelling through their logic gates.[2] Due to the costs involved in development, 5 nm is predicted to take longer to reach market than the 2 years estimated by Moore's law.[3]

15.1.2   Technology demos

In 2006, a team of Korean researchers from the Korea Advanced Institute of Science and Technology (KAIST) and the National Nano Fab Center codeveloped a 3 nm transistor, the world's smallest nanoelectronic device based on conventional technology, called a fin field-effect transistor (FinFET).[4][5] It was the smallest transistor ever produced.

In 2008, transistors one atom thick and ten atoms wide were made by UK researchers. They were carved from graphene, predicted by some to one day oust silicon as the basis of future computing. Graphene is a material made from flat sheets of carbon in a honeycomb arrangement, and is a leading contender. A team at the University of Manchester, UK, used it to make some of the smallest transistors at this time: devices only 1 nm across that contain just a few carbon rings.[6]

In 2010, an Australian team announced that they fabricated a single functional transistor out of 7 atoms that measured 4 nm in length.[7][8][9]

In 2012, a single-atom transistor was fabricated using a phosphorus atom bound to a silicon surface (between two significantly larger electrodes).[10] This transistor could be said to be a 180 picometer transistor, the Van der Waals radius of a phosphorus atom; though its covalent radius bound to silicon is likely smaller.[11] Making transistors smaller than this will require either using elements with smaller atomic radii, or using subatomic particles—like electrons or protons—as functional transistors.

In 2015 IMEC and Cadence had fabricated 5 nm test chips. The fabricated test chips are not fully functional devices but rather are to evaluate patterning of interconnect layers.[12][13]

15.1.3   Commercialization

Although Intel has not yet divulged any certain plans to manufacturers or retailers, their 2009 roadmap projected an end-user release by approximately 2020.[14][15]

15.2   References

[1] "End of Moore's Law: It's not just about physics". *CNET*. August 28, 2013.

[2] Pirzada, Usman. "Intel ISSCC: 14nm all figured out, 10nm is on track, Moores Law still alive and kicking". *WCCF Tech*. Retrieved 2015-07-02.

[3] "End of Moore's Law: It's not just about physics". *CNET*. August 28, 2013.

[4] Still Room at the Bottom.(nanometer transistor developed by Yang-kyu Choi from the Korea Advanced Institute of Science and Technology)

[5] Lee, Hyunjin; et al. (2006). "Sub-5nm All-Around Gate FinFET for Ultimate Scaling". *Symposium on VLSI Technology, 2006*: 58–59. doi:10.1109/VLSIT.2006.1705215.

[6] Atom-thick material runs rings around silicon

[7] Fuechsle, Martin; et al. (2010). "Spectroscopy of few-electron single-crystal silicon quantum dots". *Nature Nanotechnology* **5** (7): 502–505. doi:10.1038/nnano.2010.95.

[8] Ng, Jansen (May 24, 2010). "Researchers Create Seven Atom Transistor, Working on Quantum Computer". *Daily Tech*.

[9] Beale, Bob (May 24, 2010). "Quantum leap: World's smallest transistor built with just 7 atoms". *Phys.Org*.

[10] Fuechsle, M.; Miwa, J. A.; Mahapatra, S.; Ryu, H.; Lee, S.; Warschkow, O.; Hollenberg, L. C. L.; Klimeck, G.; Simmons, M. Y. (2012). "A single-atom transistor". *Nature Nanotechnology* **7** (4): 242. doi:10.1038/nnano.2012.21.

[11] "Team designs world's smallest transistor". Retrieved 28 May 2013.

[12] "IMEC and Cadence Disclose 5nm Test Chip". Retrieved 25 Nov 2015.

[13] "The Roadmap to 5nm: Convergence of Many Solutions Needed". Retrieved 25 Nov 2015.

[14] "Intel Outlines Process Technology Roadmap". Xbit. 2009-08-22.

[15] "□□□□□□32nm□□□□□□□□□□□□□□□□□□□□□" [Intel touts steady rise of 32nm processors] (in Japanese). PC Watch. 2009-08-21.

Chapter 16

600 nanometer

This article is about semiconductor manufacturing. For 600nm wavelength light, see Orange (colour).

The **600 nanometer** (**600 nm**) process refers to the level of semiconductor process technology that was reached in the 1994–1995 timeframe, by most leading semiconductor companies, like Intel and IBM.

16.1   Products featuring 0.6 μm manufacturing process

- Intel 80486DX4 CPU launched in 1994 was manufactured using this process.

- IBM/Motorola PowerPC 601, the first PowerPC chip, was produced in 0.6 μm.

- Intel Pentium CPUs at 75 MHz, 90 MHz and 100 MHz were also manufactured using this process.

Chapter 17

65 nanometer

The **65 nanometer** (**65 nm**) process is advanced lithographic node used in volume CMOS semiconductor fabrication. Printed linewidths (i.e., transistor gate lengths) can reach as low as 25 nm on a nominally 65 nm process, while the pitch between two lines may be greater than 130 nm.[1] For comparison, cellular ribosomes are about 20 nm end-to-end. A crystal of bulk silicon has a lattice constant of 0.543 nm, so such transistors are on the order of 100 atoms across. By September 2007, Intel, AMD, IBM, UMC, Chartered and TSMC were producing 65 nm chips.

While feature sizes may be drawn as 65 nm or less, the wavelengths of light used for lithography are 193 nm and 248 nm. Fabrication of sub-wavelength features requires special imaging technologies, such as optical proximity correction and phase-shifting masks. The cost of these techniques adds substantially to the cost of manufacturing sub-wavelength semiconductor products, with the cost increasing exponentially with each advancing technology node. Furthermore, these costs are multiplied by an increasing number of mask layers that must be printed at the minimum pitch, and the reduction in yield from printing so many layers at the cutting edge of the technology. For new integrated circuit designs, this factors into the costs of prototyping and production.

Gate thickness, another important dimension, is reduced to as little as 1.2 nm (Intel). Only a few atoms insulate the "switch" part of the transistor, causing charge to flow through it. This undesired effect, *leakage*, is caused by quantum tunneling. The new chemistry of high-k gate dielectrics must be combined with existing techniques including substrate bias and multiple threshold voltages to prevent leakage from prohibitively consuming power.

IEDM papers from Intel in 2002, 2004, and 2005 illustrate the industry trend that the transistor sizes can no longer scale along with the rest of the feature dimensions (gate width only changed from 220 nm to 210 nm going from 90 nm to 65 nm technologies). However, the interconnects (metal and poly pitch) continue to shrink, thus reducing chip area and chip cost, as well as shortening the distance between transistors, leading to higher performance devices of greater complexity when compared with earlier nodes.

17.1   Example: Fujitsu 65 nm process[2][3]

- Gate length: 30 nm (high-performance) to 50 nm (low-power)

- Core voltage: 1.0 V

- 11 Cu interconnect layers using nano-clustering silica as ultralow k dielectric (k=2.25)

- Metal 1 pitch: 180 nm

- Nickel silicide source/drain

- Gate oxide thickness: 1.9 nm (n), 2.1 nm (p)

There are actually two versions of the process: CS200, focusing on high performance, and CS200A, focusing on low power.

17.2   Processors using 65 nm manufacturing technology

- Intel Pentium 4 (Cedar Mill) – 2006-01-16

- Intel Pentium D 900-series – 2006-01-16

- Intel Celeron D (Cedar Mill cores) – 2006-05-28

- Intel Core – 2006-01-05

- Intel Core 2 – 2006-07-27

- Intel Xeon (Sossaman) – 2006-03-14

- AMD Athlon 64 series (starting from Lima) – 2007-02-20

- AMD Turion 64 X2 series (starting from Tyler)- 2007-05-07

- AMD Phenom series

- IBM's Cell Processor – PlayStation 3 – 2007-11-17

- IBM's z10

- Microsoft Xbox 360 "Falcon" CPU – 2007–09

- Microsoft Xbox 360 "Opus" CPU – 2008

- Microsoft Xbox 360 "Jasper" CPU – 2008–10

- Microsoft Xbox 360 "Jasper" GPU – 2008–10

- Sun UltraSPARC T2 – 2007–10

- AMD Turion Ultra – 2008-06[4]

- TI OMAP 3 Family[5] – 2008-02

- VIA Nano – 2008-05

- Loongson – 2009

- NVIDIA GeForce 8800GT GPU - 2007

- Nikon Expeed 2 - 2010

17.3   References

[1] 2006 industry roadmap, Table 40a

[2] link to press release

[3] link to presentation

[4] TG Daily – AMD preps 65 nm Turion X2 processors

[5] http://focus.ti.com/pdfs/wtbu/ti_omap3family.pdf

General

- "Intel to cut Prescott leakage by 75% at 65nm". The Register. August 31, 2004. Retrieved 2007-08-25.

- Engineering Sample of the "Yonah" core Pentium M, IDF Spring 2005, ExtremeTech

- "AMD's 65 nano silicon ready to roll". The Inquirer. September 2, 2005. Retrieved 2007-08-25.

Chapter 18

7 nanometer

In semiconductor manufacturing, the International Technology Roadmap for Semiconductors defines the **7 nanometer** (**7 nm**) node as the technology node following the 10 nm node.

Single transistor 7 nm scale devices were first produced in the early 2000s – as of 2015 commercial production of 7 nm chips is at a development stage.

18.1 History

18.1.1 Technology demos

In 2002, IBM produced a 6 nm transistor.[1]

In 2003, NEC produced a 5 nm transistor.[2]

In 2012, IBM produced a sub-10 nm carbon nanotube transistor that outperformed silicon on speed and power.[3] "The superior low-voltage performance of the sub-10 nm CNT transistor proves the viability of nanotubes for consideration in future aggressively scaled transistor technologies," according to the abstract of the paper in *Nano Letters*.[4]

In July 2015, IBM announced that they had built the first functional transistors with 7 nm technology, using a silicon-germanium process.[5][6]

18.1.2 Expected commercialisation and technologies

Although Intel has not yet divulged any certain plans to manufacturers or retailers, it has already stated that it would no longer use silicon at this node.[7] A possible replacement material for silicon would be indium gallium arsenide (InGaAs).[8]

In April 2015, TSMC announced that 10 nm production would begin in 2016, followed by 7 nm production in 2017.[9]

18.2 References

[1] IBM claims world's smallest silicon transistor

[2] NEC test-produces world's smallest transistor.

[3] "IBM: Tiny carbon nanotube transistor outshines silicon". *Cnet.com*. January 30, 2012.

[4] Franklin, Aaron D.; et al. (2012). "Sub-10 nm Carbon Nanotube Transistor". *Nano Letters* **12** (2): 758–762. doi:10.1021/nl203701g.

[5] IBM Research builds functional 7nm processor

[6] IBM Discloses Working Version of a Much Higher-Capacity Chip - NYTimes.com

[7] "ISSCC 2015: Intel 10 nm Last Silicon Node". Android Authority.

[8] Intel forges ahead to 10nm, will move away from silicon at 7nm. Feb 2015

[9] TSMC launching 10 nm FinFet Process in 2016, followed by 7 nm in 2017

18.3   External links

- Summary technology trend targets lists the 2013 predictions of the 7nm node characteristics.

Chapter 19

800 nanometer

The **800 nanometer** (**800 nm**) process refers to the level of semiconductor process technology that was reached in the 1989–1990 timeframe, by most leading semiconductor companies, like Intel and IBM.

19.1 Products featuring 0.8 μm manufacturing process

- Intel 80486 CPU launched in 1989 was manufactured using this process.

- microSPARC I launched in 1992

- First Intel P5 Pentium CPUs at 60 MHz and 66 MHz launched in 1993

Chapter 20

90 nanometer

This article is about semiconductor manufacturing. For the spaceport with FAA LID code of 90NM, see Spaceport America.

The **90 nanometer** (**90 nm**) process refers to the level of CMOS process technology that was reached in the 2004–2005 timeframe, by most leading semiconductor companies, like Intel, AMD, Infineon, Texas Instruments, IBM, and TSMC.

The origin of the 90 nm value is historical, as it reflects a trend of 70% scaling every 2–3 years. The naming is formally determined by the International Technology Roadmap for Semiconductors (ITRS).

The 193 nm wavelength was introduced by many (but not all) companies for lithography of critical layers mainly during the 90 nm node. Yield issues associated with this transition (due to the use of new photoresists) were reflected in the high costs associated with this transition.

Even more significantly, the 300 mm wafer size became mainstream at the 90 nm node. The previous wafer size was 200 mm diameter.

20.1 Example: Elpida 90 nm DDR2 SDRAM process[1]

- Use of 300 mm wafer size

- Use of KrF (248 nm) lithography with optical proximity correction

- 512 Mbit

- 1.8 V operation

- Derivative of earlier 110 nm and 100 nm processes

20.2 Processors using 90 nm process technology

- IBM PowerPC G5 970FX - 2004

- IBM PowerPC G5 970MP - 2005

- IBM PowerPC G5 970GX - 2005

- IBM "Waternoose" Xbox 360 Processor - 2005

- IBM/Sony/ Toshiba Cell Processor - 2005

33

- Intel Pentium 4 Prescott - 2004-02
- Intel Celeron D Prescott-256 - 2004-05
- Intel Pentium M Dothan - 2004-05
- Intel Celeron M Dothan−1024 - 2004-08
- Intel Xeon Nocona, Irwindale, Cranford, Potomac, Paxville - 2004-06
- Intel Pentium D Smithfield - 2005-05
- AMD Athlon 64 Winchester, Venice, San Diego, Orleans - 2004-10
- AMD Athlon 64 X2 Manchester, Toledo, Windsor - 2005-05
- AMD Sempron Palermo and Manila - 2004-08
- AMD Turion 64 Lancaster and Richmond - 2005-03
- NVIDIA GeForce 8800 GTS (G80) - 2006
- AMD Turion 64 X2 Taylor and Trinidad - 2006-05
- AMD Opteron Venus, Troy, and Athens - 2005-08
- AMD Dual-core Opteron Denmark, Italy, Egypt, Santa Ana, and Santa Rosa
- VIA C7 - 2005-05
- Loongson (Godson) 2E STLS2E02 - 2007-04
- Loongson (Godson) 2F STLS2F02 - 2008-07
- MCST-4R - 2010-12
- Elbrus-2S+ - 2011-11

20.3   See also

- Photolithography

20.4   References

[1] Elpida's presentation at Via Technology Forum 2005 and Elpida 2005 Annual Report

20.5   External links

- PC World Review
- Review ITworld
- AMD
- Fujitsu
- Intel
- August, 2002 release by Intel

Chapter 21

List of semiconductor scale examples

21.1 Products featuring 10 μm manufacturing process

- Intel 4004 CPU launched in 1971

- Intel 8008 CPU launched in 1972

- MOS Technology 6502 1 MHz CPU launched in 1975 (8 μm)

21.2 Products featuring 3 μm manufacturing process

- Intel 8085 CPU launched in 1975

- Intel 8088 CPU launched in 1979

- Motorola 68000 8 MHz CPU launched in 1979 (3.5 μm)

21.3 Products featuring 1.5 μm manufacturing process

- Intel 80286 CPU launched in 1982

21.4 Products featuring 1.0 μm manufacturing process

- Intel 80386 CPU launched in 1985

21.5 Products featuring 0.8 μm manufacturing process

- Intel 486 CPU launched in 1989

- microSPARC I launched in 1992

- First Intel P5 Pentium CPUs at 60 MHz and 66 MHz launched in 1993

21.6   Products featuring 0.6 μm manufacturing process

- Intel 80486DX4 CPU launched in 1994

- IBM/Motorola PowerPC 601, the first PowerPC chip, was produced in 0.6 μm.

- Intel Pentium CPUs at 75 MHz, 90 MHz and 100 MHz

21.7   Products featuring 350 nm manufacturing process

- Intel Pentium Pro (1995), Pentium (P54CS, 1995), and initial Pentium II CPUs (Klamath, 1997)

- AMD K5 (1996) and original AMD K6 (Model 6, 1997) CPUs.

- NEC VR4300, used in the Nintendo 64 game console

- Parallax Propeller, 8 core microcontroller[1]

21.8   Products featuring 250 nm manufacturing process

- DEC Alpha 21264A, which was made commercially available in 1999

- AMD K6-2 *Chomper* and *Chomper Extended*. Chomper was released on May 28, 1998

- AMD K6-III "Sharptooth" used 250 nm

- Mobile Pentium MMX *Tillamook*, released in August 1997

- Pentium II *Deschutes*

- Pentium III *Katmai*

- Dreamcast CPU and GPU

- Initial PlayStation 2's Emotion Engine CPU

21.9   Processors using 180 nm manufacturing technology

- Intel Coppermine E- October, 1999

- Intel Celeron (Willamette) - May, 2002

- Motorola PowerPC 7445 and 7455 (Apollo 6) - January, 2002

21.10   Processors using 130 nm manufacturing technology

- Motorola PowerPC 7447 and 7457 2002

- IBM Gekko (Nintendo GameCube)

- IBM PowerPC G5 970 - October 2002 - June 2003

- Intel Pentium III Tualatin and Coppermine - 2001-04

- Intel Celeron Tualatin−256 - 2001-10-02

- Intel Pentium M Banias - 2003-03-12

- Intel Pentium 4 Northwood- 2002-01-07

- Intel Celeron Northwood-128 - 2002-09-18

- Intel Xeon Prestonia and Gallatin - 2002-02-25

- VIA C3 - 2001

- AMD Athlon XP Thoroughbred, Thorton, and Barton

- AMD Athlon MP Thoroughbred - 2002-08-27

- AMD Athlon XP-M Thoroughbred, Barton, and Dublin

- AMD Duron Applebred - 2003-08-21

- AMD K7 Sempron Thoroughbred-B, Thorton, and Barton - 2004-07-28

- AMD K8 Sempron Paris - 2004-07-28

- AMD Athlon 64 Clawhammer and Newcastle - 2003-09-23

- AMD Opteron Sledgehammer - 2003-06-30

- Elbrus 2000 1891ВМ4Я (1891VM4YA) - 2008-04-27

- MCST-R500S 1891ВМ3 - 2008-07-27

- Vortex 86SX -

21.11   Chips using 90 nm manufacturing technology

- Elpida 90 nm DDR2 SDRAM process

- IBM PowerPC G5 970FX - 2004

- IBM PowerPC G5 970MP - 2005

- IBM PowerPC G5 970GX - 2005

- IBM "Waternoose" Xbox 360 Processor - 2005

- IBM/Sony/ Toshiba Cell Processor - 2005

- Intel Pentium 4 Prescott - 2004-02

- Intel Celeron D Prescott-256 - 2004-05

- Intel Pentium M Dothan - 2004-05

- Intel Celeron M Dothan−1024 - 2004-08

- Intel Xeon Nocona, Irwindale, Cranford, Potomac, Paxville - 2004-06

- Intel Pentium D Smithfield - 2005-05

- AMD Athlon 64 Winchester, Venice, San Diego, Orleans - 2004-10

- AMD Athlon 64 X2 Manchester, Toledo, Windsor - 2005-05

- AMD Sempron Palermo and Manila - 2004-08

- AMD Turion 64 Lancaster and Richmond - 2005-03

- AMD Turion 64 X2 Taylor and Trinidad - 2006-05

- AMD Opteron Venus, Troy, and Athens - 2005-08

- AMD Dual-core Opteron Denmark, Italy, Egypt, Santa Ana, and Santa Rosa

- VIA C7 - 2005-05

- Loongson (Godson) 2E STLS2E02 - 2007-04

- Loongson (Godson) 2F STLS2F02 - 2008-07

- MCST-4R - 2010-12

- Elbrus-2C+ - 2011-11

21.12   Processors using 65 nm manufacturing technology

- Intel Pentium 4 (Cedar Mill) – 2006-01-16

- Intel Pentium D 900-series – 2006-01-16

- Intel Celeron D (Cedar Mill cores) – 2006-05-28

- Intel Core – 2006-01-05

- Intel Core 2 – 2006-07-27

- Intel Xeon (Sossaman) – 2006-03-14

- AMD Athlon 64 series (starting from Lima) – 2007-02-20

- AMD Turion 64 X2 series (starting from Tyler)- 2007-05-07

- AMD Phenom series

- IBM's Cell Processor – PlayStation 3 – 2007-11-17

- IBM's z10

- Microsoft Xbox 360 "Falcon" CPU – 2007–09

- Microsoft Xbox 360 "Opus" CPU – 2008

- Microsoft Xbox 360 "Jasper" CPU – 2008–10

- Microsoft Xbox 360 "Jasper" GPU – 2008–10

- Sun UltraSPARC T2 – 2007–10

- AMD Turion Ultra – 2008-06[2]

- TI OMAP 3 Family[3] – 2008-02

- VIA Nano – 2008-05

- Loongson – 2009

- NVIDIA GeForce 8800GT GPU - 2007

21.13   Processors using 45 nm technology

- Matsushita has released the 45 nm Uniphier.

- Wolfdale, Yorkfield, Yorkfield XE and Penryn are current Intel cores sold under the Core 2 brand.

- Intel Core i7 series processors, i5 750 (Lynnfield and Clarksfield)

- Pentium Dual-Core Wolfdale-3M are current Intel mainstream dual core sold under the Pentium brand.

- Diamondville, Pineview are current Intel cores with Hyper-Threading sold under the Intel Atom brand.

- AMD Deneb (Phenom II) and Shanghai (Opteron) Quad-Core Processors, Regor (Athlon II) dual core processors , Caspian (Turion II) mobile dual core processors.

- AMD(Phenom II) "Thuban" Six-Core Processor (1055T)

- Xenon in the Xbox 360 S model.

- Cell Broadband Engine in PlayStation 3 Slim model – September 2009.

- Samsung S5PC110, as known as *Hummingbird*.

- Texas Instruments OMAP 36xx.

- IBM POWER7 and z196

- Fujitsu SPARC64 VIIIfx series

- Espresso (microprocessor) Wii U CPU

21.14   Chips using 32 nm technology

- Intel Core i3 and i5 processors, released in January 2010[4]

- Intel 6-core processor, codenamed Gulftown[5]

- Intel i7-970, was released in late July 2010, priced at approximately US$900

- AMD FX Series processors, codenamed Zambezi and based on AMD's Bulldozer architecture, were released in October 2011. The technology utilised a 32 nm SOI process, two CPU cores per module, and up to four modules, ranging from a quad-core design costing approximately US$130 to a $280 eight-core design.

- Ambarella Inc. announced the availability of the A7L system-on-a-chip circuit for digital still cameras, providing 1080p60 high-definition video capabilities in September 2011[6]

21.15   Chips using 22 nm technology

- Toshiba announced that it was shipping 24 nm flash memory NAND devices on August 31, 2010.[7]

- Hynix Semiconductor announced that it could produce a 26 nm flash chip with 64 Gb capacity; Intel Corp. and Micron Technology had by then already developed the technology themselves. Announced in 2010.[8]

- Intel Core i7 and Intel Core i5 processors based on Intel's Ivy Bridge 22 nm technology for series 7 chip-sets went on sale worldwide on April 23, 2012.[9]

21.16   Chips using 14 nm technology

- Intel Core i7 and Intel Core i5 processors based on Intel's Broadwell 14 nm technology was launched in January 2015.[10]

21.17   References

[1] http://forums.parallax.com/showthread.php?130327-Propeller-I-semiconductor-process-technology-Is-it-350nm-or-180nm

[2] TG Daily – AMD preps 65 nm Turion X2 processors

[3] http://focus.ti.com/pdfs/wtbu/ti_omap3family.pdf

[4] "Intel Debuts 32-NM Westmere Desktop Processors". InformationWeek, 7 January 2010. Retrieved 2011-12-17.

[5] Sal Cangeloso (February 4, 2010). "Intel's 6-core 32nm processors arriving soon". *Geek.com*. Retrieved November 11, 2011.

[6] "Ambarella A7L Enables the Next Generation of Digital Still Cameras with 1080p60 Fluid Motion Video". *News release*. September 26, 2011. Retrieved November 11, 2011.

[7] Toshiba launches 24nm process NAND flash memory

[8] Article reporting Hynix 26 nm technology announcement

[9] Intel launches Ivy Bridge...

[10] EETimes Intel Rolls 14nm Broadwell in Vegas

21.18 Text and image sources, contributors, and licenses

21.18.1 Text

- **1 µm process** *Source:* https://en.wikipedia.org/wiki/1_%C2%B5m_process?oldid=681972024 *Contributors:* D6, Shanes, Paul1337, Stuartyeates, GregorB, Chobot, Jaraalbe, Avicennasis, Darin-0, Jeepday, Steven Crossin, Addbot, Dawynn, Abduallah mohammed, Luckas-bot, Materialscientist, WebCiteBOT, Erik9bot, DrilBot, A conundrum, John of Reading, ZéroBot, Hlm Z., ChrisGualtieri, Hmainsbot1, Parlinone and Anonymous: 3

- **1.5 µm process** *Source:* https://en.wikipedia.org/wiki/1.5_%C2%B5m_process?oldid=678898354 *Contributors:* Heron, D6, Paul1337, Stuartyeates, Chobot, Jaraalbe, Colinstu, Darin-0, Jeepday, Steven Crossin, PipepBot, Addbot, Dawynn, Abduallah mohammed, Yobot, AnomieBOT, Materialscientist, WebCiteBOT, Erik9bot, RedBot, A conundrum, John of Reading, WikitanvirBot, ZéroBot, Brustopher, ChrisGualtieri, Hmainsbot1, Parlinone and Anonymous: 2

- **10 µm process** *Source:* https://en.wikipedia.org/wiki/10_%C2%B5m_process?oldid=678897895 *Contributors:* Heron, D6, Axl, Paul1337, Stuartyeates, Allen3, Chobot, Jaraalbe, Colinstu, Hmains, EMINESCU, Magioladitis, Darin-0, Jeepday, Potatoswatter, Addbot, Dawynn, Abduallah mohammed, Yobot, Materialscientist, WebCiteBOT, Erik9bot, Thehelpfulbot, WikitanvirBot, Hlm Z., AdventurousSquirrel, ChrisGualtieri, Parlinone and Anonymous: 4

- **6 µm process** *Source:* https://en.wikipedia.org/wiki/6_%C2%B5m_process?oldid=678898121 *Contributors:* Heron, Imzadi1979, William1000000 and Parlinone

- **10 nanometer** *Source:* https://en.wikipedia.org/wiki/10_nanometer?oldid=699334099 *Contributors:* DmitryKo, Rich Farmbrough, FT2, Bender235, Ronark, Crosbiesmith, Marasmusine, JFG, Rjwilmsi, Ysangkok, Wjfox2005, David R. Ingham, Alpertron, SmackBot, Colinstu, Otus, Guiding light, Frap, Pizzahut2, Iliev, Joffeloff, CRGreathouse, Ashankar, Widefox, Poshzombie, Registrar, Markhahn, Schmloof, Rod57, Cometstyles, TXiKiBoT, Lordvolton, Nacarlson, Coinmanj, Rreagan007, Addbot, Dawynn, Smeagol 17, Legobot, Yobot, AnomieBOT, Citation bot, WebCiteBOT, Trappist the monk, EmausBot, PlatoCantRepent, Beatnik8983, Mmeijeri, Ausernamenottaken, Cogiati, H3llBot, Violettsureme, BG19bot, Virtualerian, Ushio01, Comp.arch, Stamptrader, Dark Mistress, Jacob Gotts, Golddancer, EnigmaLord515, Xiiophen and Anonymous: 40

- **130 nanometer** *Source:* https://en.wikipedia.org/wiki/130_nanometer?oldid=680401278 *Contributors:* Bender235, Matt Britt, Rabarberski, Paul1337, Stuartyeates, Chsf, Chobot, Jaraalbe, Elonka, SmackEater, O keyes, Pgk1, Grapetonix, Kvng, Fernvale, Widefox, Tracer9999, Darin-0, Dstary, Rreagan007, Addbot, Dawynn, LaaknorBot, Borisich, Materialscientist, FaleBot, Gnomsovet, MoreNet, ZéroBot, Yiosie2356 and Anonymous: 15

- **14 nanometer** *Source:* https://en.wikipedia.org/wiki/14_nanometer?oldid=696694025 *Contributors:* Karolkalna, Micru, Rchandra, AlistairMcMillan, Bumm13, Rich Farmbrough, Bender235, Matt Britt, Paul1337, Crosbiesmith, Ysangkok, Jaraalbe, Nehalem, Cwlq, Ergzay, SmackBot, Carbon-16, Guiding light, Frap, Ohconfucius, Fernvale, CmdrObot, CBM, Icek~enwiki, W.F.Galway, Dawkeye, Arch dude, ShaunL, Ahvetm, Darin-0, Falcon866, Potatoswatter, Ajfweb, TXiKiBoT, Lordvolton, Qaywsxedc, Neparis, Reinderien, Dcook32p, Hairmerchant, Scalziand, DumZiBoT, Rreagan007, Airplaneman, Addbot, Racecar56, Smeagol 17, Yobot, AnomieBOT, Diadistis, Xmagic5589, WebCiteBOT, Trappist the monk, Oliver H, Jwolla, MegaSloth, EmausBot, Alguemimportante, Dewritech, Hanjifi, Filterlol, Michaelmas1957, Imperi, BG19bot, PC-XT, DouGami, Comp.arch, Makkachin, Gvolker, Parlinone, Otistd, Xiiophen and Anonymous: 55

- **180 nanometer** *Source:* https://en.wikipedia.org/wiki/180_nanometer?oldid=628563592 *Contributors:* David Latapie, Rich Farmbrough, Matt Britt, Rabarberski, Paul1337, Stuartyeates, Christopher Thomas, Chobot, Jaraalbe, Elonka, SmackEater, O keyes, Iliev, MaximusBrood, Fernvale, Baiji, Deusnoctum, Tracer9999, Darin-0, ARTE, Rreagan007, Addbot, Dawynn, Materialscientist, Erik9bot, ZéroBot and Anonymous: 13

- **22 nanometer** *Source:* https://en.wikipedia.org/wiki/22_nanometer?oldid=695470619 *Contributors:* FlyByPC, Nurg, Rich Farmbrough, Hydrox, Bender235, Matt Britt, Lawpjc, Atlant, Irdepesca572, Paul1337, Suruena, Crosbiesmith, Bob A, Zayani, Jaraalbe, Nehalem, Bspoka, Vuvar1, TheDoober, SmackBot, Elonka, Henriok, Des1974, Guiding light, Cybercobra, Pizzahut2, Shirifan, Mtodorov 69, Fernvale, Widefox, Jace1982~enwiki, David Craft, Darin-0, ARTE, Bmws88, Wskitche, Rreagan007, Addbot, Mortense, Borisich, Smeagol 17, Spacy73, Yobot, WANGUN Styleways, PluniAlmoni, Materialscientist, Wimchatta, Uusijani, FrescoBot, Mitizhi, Orgio89, Pinethicket, Diannaa, EmausBot, Mhinterseher, ZéroBot, Cogiati, AUN4, 2001:db8, Galots, BattyBot, Paulreynolds245, Comp.arch and Anonymous: 50

- **250 nanometer** *Source:* https://en.wikipedia.org/wiki/250_nanometer?oldid=545289449 *Contributors:* D6, Rich Farmbrough, Alistair1978, Denniss, Paul1337, Stuartyeates, Chobot, Jaraalbe, TJ Spyke, Cheztir, Ciao 90, Darin-0, Falcon866, Forkazoo, PipepBot, Rreagan007, Addbot, Dawynn, Abduallah mohammed, Materialscientist, Erik9bot, ZéroBot and Anonymous: 8

- **3 µm process** *Source:* https://en.wikipedia.org/wiki/3_%C2%B5m_process?oldid=680350435 *Contributors:* Heron, D6, Rich Farmbrough, Paul1337, Stuartyeates, MZMcBride, Chobot, Jaraalbe, Colinstu, Hmains, Novous, Widefox, Magioladitis, Darin-0, Reedy Bot, Potatoswatter, Addbot, Dawynn, TutterMouse, Abduallah mohammed, Yobot, Materialscientist, Erik9bot, DrilBot, RedBot, John of Reading, WikitanvirBot, Hlm Z., ChrisGualtieri, Parlinone and Anonymous: 8

- **32 nanometer** *Source:* https://en.wikipedia.org/wiki/32_nanometer?oldid=693899172 *Contributors:* Palfrey, Gracefool, Bender235, Nekochan, Matt Britt, KBi, Alansohn, Paul1337, LOL, Compotatoj, Kickboy, Chobot, Jaraalbe, Cwlq, Anomalocaris, Everyguy, Superluser, Caco de vidro, NetRolller 3D, SmackBot, Chris the speller, Bluebot, Guiding light, Frap, Cybercobra, Iliev, Amendt, Fernvale, CmdrObot, W.F.Galway, Widefox, Arch dude, Inspector 2211, Magioladitis, Tracer9999, Pvosta, Darin-0, ARTE, SieBot, Int21h, Brickwall04, Rreagan007, Addbot, Kmeisterling, Download, Yobot, Materialscientist, Gerby123, Lonaowna, W Nowicki, Trappist the monk, Werieth, Michaelmas1957, Deepon, 🔲🔲🔲🔲🔲🔲, Ices2Csharp, MostlyCarbon and Anonymous: 56

- **350 nanometer** *Source:* https://en.wikipedia.org/wiki/350_nanometer?oldid=680767434 *Contributors:* D6, Rich Farmbrough, Paul1337, Stuartyeates, Ysangkok, Chobot, Jaraalbe, Swaaye, Widefox, Morenita~enwiki, Darin-0, Rilak, Dthomsen8, Reagan007, Addbot, Dawynn, Abduallah mohammed, Luckas-bot, Materialscientist, Erik9bot, Arndbergmann, ZéroBot, BattyBot and Anonymous: 6

- **45 nanometer** *Source:* https://en.wikipedia.org/wiki/45_nanometer?oldid=696320345 *Contributors:* Dhart, Altenmann, Jondel, Asparagus, Gracefool, Bobblewik, Gunkyman, Alistair1978, Bender235, Triona, Matt Britt, Guy Harris, Paul1337, LOL, Rjwilmsi, Kickboy, Chobot, Jaraalbe, Nehalem, YurikBot, The1physicist, Code65536, Syrthiss, Everyguy, Chris S, G-smooth2k, NetRolller 3D, AnOddName, Bluebot, Droll, BBCWatcher, Guiding light, Steveo1544, Nicklinn, Nasz, Ohconfucius, Scientizzle, Joffeloff, CapitalR, Fernvale, Ruslik0, Anonymous 198736, Big Kahoona, Geekosaurus, Headbomb, Segfault87, Lithpiperpilot, Widefox, DuncanHill, TAnthony, Magioladitis, Ciao 90, Debollweevil, Gwern, Hemidemisemiquaver, ShaunL, Darin-0, Potatoswatter, ACSE, James Callahan, WikipedianYknOK, Vinnivince, Android Mouse, Rilak, Bender2k14, Toadster-CA, Rreagan007, Airplaneman, Addbot, Fireaxe888, Materialscientist, DSisyphBot, Gumok, WikiD71, Arndbergmann, PigFlu Oink, Trappist the monk, DASHBot, EmausBot, Dewritech, Werieth, ZéroBot, Crown Prince, Muon, Electriccatfish2, BattyBot, Stamptrader, Sofia Koutsouveli and Anonymous: 96

- **5 nanometer** *Source:* https://en.wikipedia.org/wiki/5_nanometer?oldid=700062848 *Contributors:* Rainer Wasserfuhr~enwiki, Graeme Bartlett, SECProto, Xezbeth, Bender235, Richard Arthur Norton (1958-), Colinstu, Hibernian, Nbarth, Egsan Bacon, Guiding light, Pizzahut2, Twinpinesmall, Kvng, FredWallace18@yahoo.com, Widefox, Jobu0101, Manchurian candidate, Mikebar, Phil Bridger, Rreagan007, OlEnglish, AnomieBOT, Quebec99, FrescoBot, Trappist the monk, Yunshui, Stealthpaladin, Shookees, Mmoecke, ZéroBot, SkywalkerPL, Julesmazur, Danim, Technical 13, BG19bot, Virtualerian, Kizar, Minitech.me, Fraulein451, ChrisGualtieri, Dexbot, Wethar555, Comp.arch, Scarabola, Polemic Thoughts, BethNaught, Jacob Gotts, ScrapIronIV, Xiiophen and Anonymous: 36

- **600 nanometer** *Source:* https://en.wikipedia.org/wiki/600_nanometer?oldid=680350284 *Contributors:* Mboverload, D6, Rich Farmbrough, Paul1337, Stuartyeates, GregorB, Chobot, Jaraalbe, Hmains, Widefox, Darin-0, Rreagan007, Addbot, Dawynn, Abduallah mohammed, Luckas-bot, Materialscientist, Erik9bot, Arndbergmann, ZéroBot and Anonymous: 5

- **65 nanometer** *Source:* https://en.wikipedia.org/wiki/65_nanometer?oldid=680349733 *Contributors:* Jerec, Altenmann, Jondel, Fudoreaper, Dratman, Gracefool, Fastred, Sam Hocevar, Frankchn, Rich Farmbrough, Alistair1978, Josephycc, Matt Britt, Defsac, Guy Harris, Fawcett5, Paul1337, Oleg Alexandrov, Koavf, KamasamaK, SkiDragon, Kickboy, Chobot, Jaraalbe, YurikBot, The1physicist, Mipadi, Robertvan1, Chris S, Cmskog, SmackBot, Bluebot, BBCWatcher, Guiding light, Frap, Steveo1544, Ruw1090, Sloverlord, Pgk1, Iliev, Fernvale, Myscrnnm, X96lee15, Widefox, Arch dude, Magioladitis, Retroneo, Darin-0, Potatoswatter, GCFreak2, RJASE1, IO3, Reinderien, Alfonsedode, Brews ohare, DumZiBoT, AlecEspie, Rreagan007, Airplaneman, Addbot, Materialscientist, LovesMacs, Metalindustrien, ZéroBot, Juliusbaxter, Tagremover and Anonymous: 68

- **7 nanometer** *Source:* https://en.wikipedia.org/wiki/7_nanometer?oldid=697920144 *Contributors:* Graeme Bartlett, Xezbeth, Bender235, JFG, Wjfox2005, Colinstu, Guiding light, Pizzahut2, Twinpinesmall, Kvng, John a s, Rod57, 28bytes, Lordvolton, Manchurian candidate, Mikebar, Phil Bridger, Rreagan007, Skrutten84, AnomieBOT, Trappist the monk, Yunshui, JoeSperrazza, Technical 13, Virtualerian, Comp.arch, Jacob Gotts, Golddancer, Xiiophen and Anonymous: 10

- **800 nanometer** *Source:* https://en.wikipedia.org/wiki/800_nanometer?oldid=594495616 *Contributors:* D6, Rich Farmbrough, Paul1337, Stuartyeates, Chobot, Jaraalbe, Novous, Darin-0, Rreagan007, Addbot, Dawynn, Abduallah mohammed, Luckas-bot, Yobot, Materialscientist, Erik9bot, ZéroBot, Hmainsbot1 and Anonymous: 5

- **90 nanometer** *Source:* https://en.wikipedia.org/wiki/90_nanometer?oldid=680349773 *Contributors:* Heron, Bloodshedder, Altenmann, Jondel, Gracefool, Bobblewik, Zeimusu, Sam Hocevar, Jkl, ArnoldReinhold, Agoode, Duk, Matt Britt, Cohesion, Homerjay, Wendell, Paul1337, Blaxthos, Dismas, Koavf, Jevon, Kevmitch, Chobot, Jaraalbe, Wjfox2005, Aluvus, YurikBot, Cwlq, The1physicist, Bovineone, Voidxor, Chris S, Nothlit, Bluebot, Steveo1544, Sloverlord, ElementFire, A5b, Ohconfucius, Pgk1, Iliev, Fernvale, Widefox, Maximus06, Magioladitis, Tracer9999, R'n'B, Darin-0, Sulimo, Potatoswatter, STBotD, TXiKiBoT, RedAndr, Venny85, -=HyPeRzOnD=-, Rreagan007, Addbot, Dawynn, Materialscientist, Gnomsovet, FrescoBot, ZéroBot, Strak Jegan and Anonymous: 21

- **List of semiconductor scale examples** *Source:* https://en.wikipedia.org/wiki/List_of_semiconductor_scale_examples?oldid=669970198 *Contributors:* Guy Harris, Chris the speller, Electron9, Robsalmond, Just1morerifle, ChrisGualtieri and Dexbot

21.18.2 Images

- **File:4-fach-NAND-C10.JPG** *Source:* https://upload.wikimedia.org/wikipedia/commons/d/d5/4-fach-NAND-C10.JPG *License:* CC BY-SA 3.0 *Contributors:* Own work *Original artist:* Dgarte

- **File:Nuvola_apps_ksim.png** *Source:* https://upload.wikimedia.org/wikipedia/commons/8/8d/Nuvola_apps_ksim.png *License:* LGPL *Contributors:* http://icon-king.com *Original artist:* David Vignoni / ICON KING

- **File:Question_book-new.svg** *Source:* https://upload.wikimedia.org/wikipedia/en/9/99/Question_book-new.svg *License:* Cc-by-sa-3.0 *Contributors:*
  Created from scratch in Adobe Illustrator. Based on Image:Question book.png created by User:Equazcion *Original artist:*
  Tkgd2007

- **File:Rotaxane.png** *Source:* https://upload.wikimedia.org/wikipedia/commons/8/85/Rotaxane.png *License:* CC-BY-SA-3.0 *Contributors:* http://commons.wikimedia.org/wiki/Image:Rotaxane.jpg *Original artist:* <a href='//commons.wikimedia.org/wiki/User:Thingg' title='User:Thingg'>Thingg</a><a href='//co

21.18.3 Content license

- Creative Commons Attribution-Share Alike 3.0